SPACE

Planets

Robin Birch

CHELSEA CLUBHOUSE

An Imprint of Chelsea House Publishers

A Haights Cross Communications Company

Philadelphia

Chelsea Clubhouse
1974 Sproul Road, Suite 400
Broomall, PA 19008-0914

The Chelsea House world wide web address is www.chelseahouse.com

Library of Congress Cataloging-in-Publication Data

Birch, Robin.
   Planets / by Robin Birch.
   p. cm. — (Space)

   Includes index.
   Summary: Explores the nine planets in the solar system, describing their physical makeup, their moons, and their location in the solar system.

   ISBN 0-7910-6972-9
   1. Planets—Juvenile literature. [1. Planets.] I. Title. II. Series.
   QB602 .B57 2003
   523.4—dc21

                                                                          2002000040

First published in 2001 by
MACMILLAN EDUCATION AUSTRALIA PTY LTD
627 Chapel Street, South Yarra, Australia, 3141

Copyright © Robin Birch 2001
Copyright in photographs © individual photographers as credited

Edited by Carmel Heron and Louisa Kost
Cover and text design by Anne Stanhope
Illustrations by Frey Micklethwait

Printed in China

Acknowledgements
Cover photograph: The planet Saturn, courtesy of Photo Essentials.

Photographs courtesy of: Digital Vision, pp. 5, 12, 21, 26; NASA, pp. 11, 15, 22; NASA, supplied by Astrovisuals, pp. 6, 8, 13, 14, 16, 17, 19, 23, 24, 25, 27, 29; David Parer & Elizabeth Parer-Cook/Auscape, p. 10; Photo Essentials, pp. 1, 9, 20; Photolibrary.com/Photosearch International, p. 18.

While every care has been taken to trace and acknowledge copyright the publisher tenders their apologies for any accidental infringement where copyright has proved untraceable.

Contents

The Planets

Planets are huge balls of rock or gas. There are nine planets that **orbit** the Sun.

The Solar System

The Sun is a **star**. Planets, **comets**, and **asteroids** all orbit the Sun. **Moons** orbit planets. Together, all these bodies make up the solar system.

Mercury

Mercury is the closest planet to the Sun. It is a rocky planet. Deep holes called craters cover the surface.

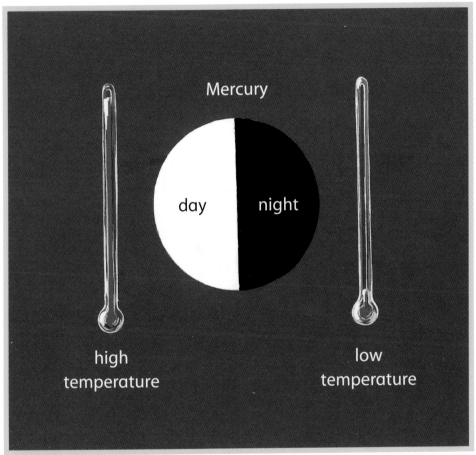

Mercury

day   night

high
temperature

low
temperature

Mercury is very hot during the day and very cold at night. The planet has almost no air. Scientists think Mercury may have water ice at its **poles**.

Venus

Venus is the second planet from the Sun. This rocky planet has mountains and craters. More than 1,000 **volcanoes** also dot the surface of Venus.

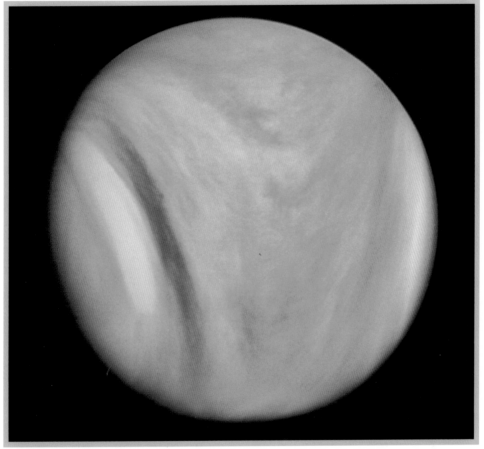

Venus is the hottest planet. Its thick **atmosphere** holds drops of **acid**. The atmosphere traps heat from the Sun and keeps the planet very hot.

Earth

Earth is the planet that we live on. It is the third planet from the Sun. Earth is made of rock. One moon orbits Earth.

Earth has water and air. Its temperatures are
not too hot or too cold. These conditions allow
Earth to support people, plants, and animals.
Earth is the only planet known to have life.

Mars

Mars is the fourth planet from the Sun. This rocky red planet has two small moons called Phobos and Deimos.

A huge **canyon** called Valles Marineris
crosses Mars. The canyon is about 2,500 miles
(4,000 kilometers) long. Olympus Mons rises
from the surface of Mars. At 16 miles
(26 kilometers) tall, it is the largest volcano
in the solar system.

Scientists sent robots to Mars. The robots took photographs of the planet's surface. The pictures showed rough land covered in rocks and dust. The daytime sky is yellow-brown.

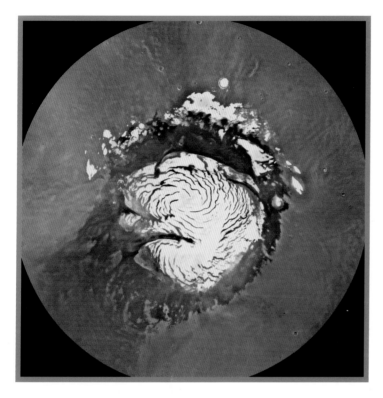

The polar ice caps on Mars grow larger during winter.

Scientists believe Mars has ice caps of water at its poles. The size of the ice caps changes with the **seasons**. When it is winter on one pole, it is summer on the other pole. The ice caps grow larger in winter and shrink in summer.

Jupiter

Jupiter is the fifth planet from the Sun. It is the largest planet in the solar system. Jupiter is a ball of gas. It probably has no solid surface. Three rings circle Jupiter. But they are too thin to be seen in photographs.

Jupiter has four large moons and 12 small
moons. Europa (pictured above) is one of the
large moons. It may have oceans of water
beneath its frozen surface.

Jupiter looks striped. It has light zones and dark belts of clouds. Strong winds blow the clouds around the planet.

Swirling clouds circle Jupiter.

Many storms rage within Jupiter's zones and belts. The Great Red Spot is a huge storm that has existed for at least 100 years. Sometimes the Great Red Spot turns brown or gray.

Saturn

Saturn is the sixth planet from the Sun. It is made of gas. Yellow and gold clouds circle Saturn.

Saturn has seven rings. They do not touch the planet. Each ring has thousands of narrow ringlets made of ice pieces. Some pieces are the size of dust. Others are as large as 10 feet (3 meters) wide.

Saturn has at least 30 moons.

This image shows Saturn and six of its moons.

Saturn's largest moon is Titan. A thick, red atmosphere always covers Titan.

Uranus

Uranus is the seventh planet from the Sun. It is a giant ball of gas and liquid. Pale blue-green clouds cover the surface of Uranus.

This photograph shows Uranus and five of its moons.

Uranus has 11 rings made of dust and ice. The 10 outer rings are dark and narrow. The inner ring is much wider. Uranus has 21 known moons that orbit the planet.

Neptune

Neptune usually is the eighth planet from the Sun. This blue planet has an icy **core** surrounded by liquid. The surface of Neptune is a thick layer of gas. Strong winds blow clouds around the planet. Dark spots in the clouds are storms. Neptune has four dark rings that circle the planet.

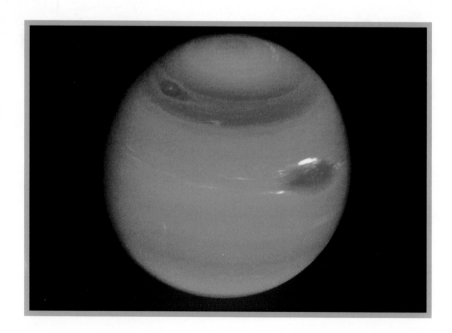

Neptune has eight moons. The two largest moons are Triton and Nereid.

Neptune's largest moon is Triton.

Pluto

Pluto usually is the farthest planet from the Sun. At times, Pluto's orbit takes the planet closer to the Sun than Neptune. Pluto is made of rock and ice.

Pluto is the coldest and smallest planet in the solar system. It is much smaller than Earth.

Pluto

Charon

The Sun

Pluto has one moon named Charon. It is half
the size of Pluto. From Pluto and Charon, the
Sun looks like a bright and faraway star.

Planet Facts

Planet	Color	Made of	Number of Moons
Mercury	gray	rock	0
Venus	gray and white	rock	0
Earth	blue and white	rock	1
Mars	red-brown	rock	2
Jupiter	red, yellow, white, brown	gas	16
Saturn	gold and yellow	gas	30
Uranus	blue-green	gas	21
Neptune	blue	gas	8
Pluto	gray	rock	1

Glossary

acid   a strong liquid that can burn other substances

asteroid   rocky objects that orbit the Sun; most asteroids are in a belt between Mars and Jupiter.

atmosphere   the mixture of gases that surround a planet

canyon   a long, deep valley with steep sides

comet   a ball of ice, frozen gases, and rock that travels through space; a comet has a tail that faces away from the Sun.

core   the center part of a planet

moon   a natural object that orbits a planet

orbit   to circle an object in space; also, the path that a planet or moon takes when circling another object.

pole   either end of a planet's axis; the axis is an imaginary line that runs through the middle of a planet; Earth has a North Pole and a South Pole.

season   a period of time each year with a certain kind of weather; the four seasons are spring, summer, autumn, and winter.

star   a large, burning ball of gas in space; a star gives off light and heat.

volcano   a hole in the surface of a planet through which hot lava and ash erupt; the lava from a volcano can harden to form a mountain.

Index